IMAGES OF ENGLAND

STAFFORDSHIRE AND BLACK COUNTRY AIRFIELDS

The Hartill monoplane at Dunstall Park, Wolverhampton, in 1910. Built in Wolverhampton by Edgar Hartill for a Dr Hands of Birmingham, it was entered for the first All-British Flying Meeting, held there in June, but failed to fly.

English Electric Lightning F.1, XG336, partially converted to F.3 standard by Boulton Paul Aircraft at their Flight Test Centre at Seighford, in 1962.

IMAGES OF ENGLAND

STAFFORDSHIRE AND BLACK COUNTRY AIRFIELDS

ALEC BREW

The History Press

First published in 1997 by Tempus Publishing Limited
Reprinted 2001

Reprinted in 2010 by
The History Press
The Mill, Brimscombe Port,
Stroud, Gloucestershire, GL5 2QG
www.thehistorypress.co.uk

ISBN 978 0 7524 0770 8

Typesetting and origination by
Tempus Publishing Limited
Printed and bound in England

Other books in The *Archive Photographs* series by Alec Brew

Albrighton and Shifnal
Codsall and Claregate
Shropshire Airfields
Tettenhall and Pattingham
Also *Boulton Paul Aircraft* by the Boulton Paul Association

Cover Pictures: Front: Bob Arnold, of the Staffordshire Aircraft Restoration Team, flying his MW Microlight, G-MNMW 'Big Prop Bob' at Dunstall Park in June 2010. Back: The Bleriot monoplane (No.5) of Julien Mamet outside his hangar 1910.

Tiger Moth G-AHLA outside the clubhouse and hangar at Walsall Airport in 1947. Tiger Moths were to be found on many Staffordshire airfields before, during and after the Second World War.

Contents

The Farman biplane of Brunneau de Laborie at the Burton-on-Trent Flying Meeting of 1910. It was held on Bass's Meadow on an island in the Trent, which was used for all flying in the Burton area for the next thirty years.

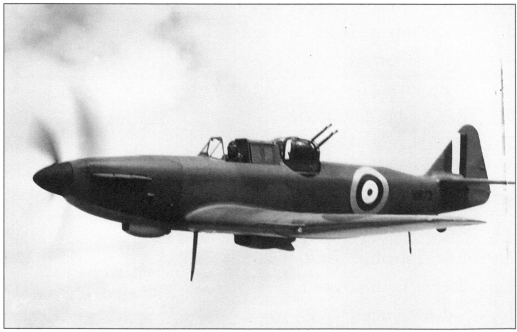

Boulton Paul Defiant N1673 on air-test from Pendeford in August 1940. This aircraft went on to serve No.264 Squadron during the Battle of Britain, No.141 Squadron during the *Blitz*, and then No.2 Air Gunners School and No.277 Air-Sea Rescue Squadron, finally being struck off charge in September 1943.

Introduction

Staffordshire has had an up and down relationship with the aeroplane. Airfields in the county have come and gone in three distinct phases, varying in number from none to the sixteen that were in use at the end of the Second World War.

Yet Staffordshire had one of the first designated airfields in the whole country, when the Racecourse Company allowed the Midland Aero Club to use Dunstall Park, Wolverhampton, as its flying field, and in June 1910, the first All-British Flying Meeting was held there - the largest gathering of aircraft to that date in this country. Bass's Meadow, Burton-on-Trent, was also used for flying meetings in 1910 and 1913, and though it was hardly a permanent airfield, it continued to be a convenient place from which barnstormers and flying circuses could fly well into the 1930s; in 1929 it was used for the first flight of the Burton-built prototype Civilian Coupe.

Dunstall Park also continued to be used for flying until just after the end of the First World War, particularly by aircraft built, and often powered by the engines of the Sunbeam Motor Co. in Wolverhampton. Two other airstrips were in use in Staffordshire during that war. Number 38 (Home Defence) Squadron used the Fern Fields at Perton, just north of Wolverhampton, as a relief landing ground, and a strip at Halford Lane, Smethwick, was used to fly the Handley Page O/400 and DH.10 bombers built by the Birmingham Carriage Company.

Although barnstormers used convenient fields like Bass's Meadow and Perton during the fifteen years after the First World War, and privately-owned light aircraft were sometimes flown from them, there were in fact no designated airfields in Staffordshire at all during this period.

In the mid 1930s, prompted by the likes of Alan Cobham, many towns and cities became rather more air-minded and civic pride determined that they should have municipal airports. Stoke, Walsall and Wolverhampton all constructed airports during the years 1934-36, at Meir, Aldridge and Pendeford respectively: in each case a grass airfield with a single small hangar and a clubhouse for a resident flying club. Wolverhampton Airport was to see rather more use than the others when the aircraft department of the Norwich company, Boulton & Paul Ltd, was sold off and became Boulton Paul Aircraft Ltd, moving to a brand new factory at Pendeford, from which 2,000 aircraft were to be built during the next twenty years.

This second flowering of aviation in the county was to be greatly expanded with the approaching Second World War and Staffordshire was to be covered with RAF training airfields. There were major training bases at Bobbington (later Halfpenny Green), Lichfield, Hixon and Wheaton Aston, with substantial satellite airfields at Seighford, Perton, and

Tatenhill. The three municipal airports were also turned over to military flying training, with smaller grass satellite airfields at Abbots Bromley, Battlestead Hill and Penkridge. Meir was also used for the flight testing of the Blenheims and Beaufighters built by the Rootes Bros.-operated Shadow Factory at Blythe Bridge, so that during the Second World War there were two aircraft manufacturers, one at each end of the county.

Finally, there were satellite strips in the grounds of the stately homes at Hoar Cross and Teddesley Park, which were used for the storage of aircraft. As a footnote to this expansion, perhaps the most important RAF base in the county was RAF Stafford, but this never had an airfield, remaining to this day one of the RAF's most important storage facilities; nevertheless aircraft have often been seen there, not least guarding the gate.

The end of the Second World War saw the second great contraction of flying in Staffordshire, with the closure of almost every RAF station. Only Lichfield and Halfpenny Green saw much post-war flying, with a significant revival at the time of the Korean War, but by the end of the 1950s these too were closed, leaving Stafford as the only operating RAF station in the county.

The three municipal airports reopened for civil flying after 1945, but without resident flying clubs, Walsall and Stoke saw limited flying before they just faded away. Wolverhampton was rather busier, with Don Everall Aviation running a thriving club, and even operating scheduled services, and there was the continued flight testing by Boulton Paul Aircraft. But by 1971, the encroaching houses on two sides saw to it that Pendeford too was closed. RAF Seighford had been reopened and operated by Boulton Paul as the Flight Test Centre for its Canberra and Lightning conversions for nine years until 1965, but then it became one of the more obscure victims of the TSR.2 cancellation.

For a short while the only operating airfield in the county was right at the southern tip, at Halfpenny Green. Reopened in 1961, against a tidal wave of local objections, as a general aviation centre, it has become a thriving little airfield serving the Black Country and surrounding areas, and has formed the prototype for the third expansion of aviation in Staffordshire as former military airfields are reopened for civilian use.

Tatenhill, near Burton-on Trent, was a convenient place for Allied Breweries to base their corporate aircraft, and has now become another busy airfield. Seighford was reopened for the second time as the home of the Staffordshire Gliding Club, as well as a base for a number of privately-owned aircraft, and both Lichfield and Penkridge have become the homes of microlight clubs so that at the time of writing, Staffordshire can boast five operational airfields. In some cases, like Battlestead Hill, Pendeford and Perton, Staffordshire airfields have disappeared completely under bricks and mortar; in others, like Abbots Bromley and Hoar Cross, agriculture has reclaimed the land.

This book tries to recall the days when aviation filled the Staffordshire skies with the wood and fabric aircraft of the pioneers, and the swarming wartime trainers which helped teach the greatest proportion of a generation's young men to fly which will probably ever be achieved: a generation of young men that learned to fly and went to war, many in aircraft built in Staffordshire, by Boulton Paul at Pendeford and the Rootes Bros. at Blythe Bridge.

Many silent, derelict buildings serve as reminders of those days, but in Wolverhampton, a more substantial tribute to Staffordshire aviators is under development. The Boulton Paul Association is in the process of opening the Staffordshire Aircraft Museum to preserve and display the whole history of aviation in the county, and to serve as a respository, within its West Midlands Aviation Archive, of documents, photographs and other artefacts, as an educational and historical archive for generations to come.

One

Dunstall Park, Wolverhampton

In 1910 one of the first designated airfields in Great Britain was established at Dunstall Park, Wolverhampton. The Midland Aero Club had been formed in September 1909, and the Racecourse Company offered Dunstall Park as its airfield. The first All-British Flying Meeting, and only the third ever, following meetings at Doncaster and Blackpool the previous year which had been dominated by French entrants, was held from 29 June 1910.

A row of wooden hangars was erected along the canal side of the course and despite awful weather, huge crowds attended. Of the seventeen British Aviator's Certificate holders then in existence, eleven flew at Dunstall Park that week and the bulk of the prize money went to Claude Grahame-White, then one of the most famous British aviators. Some of the pilots taking part had French Aviator's Certificates. Two Wolverhampton-built aircraft were entered for the meeting - the Star and Hartill monoplanes - but neither managed to fly.

The Star did take to the air the following year and flying continued at Dunstall Park until the start of the First World War, with aircraft, airships and balloons. The Sunbeam Motor Car Company began making aero-engines in Wolverhampton in 1912 and aircraft powered by them were often to be seen at Dunstall Park, before, during, and for a short while after the war ended, but as Sunbeam's brief foray into the aircraft industry faded away, the racecourse reverted solely to one horse-power displays.

INSPECTING THE FLYING MACHINES AT DUNSTALL PARK.

The crowds at Dunstall Park inspecting the aircraft in their hangars. The first is that occupied by Claude Grahame-White's Farman, the second, Cecil Grace's Short S.27, then the Farman of George Cockburn and the Lane monoplane of H.J.D. Astley.

Before the meeting Army Captain George William Patrick Dawes attempted to gain his Aviator's Certificate flying his Humber monoplane, but on 17 June he stalled and crashed. He is shown on the left surveying the wreck, which was repaired, and he gained Certificate No.17 on 26 July. He took his monoplane to India and, on 26 March 1911, was forced to make a landing on the Bombay to Baroda railway line. The Humber was then hit by an approaching goods train, but Dawes jumped clear. He had served in the British Army in the Boer War, during the First World War he commanded the Royal Flying Corps in the Balkans, and in the Second World War served in the Royal Air Force.

THE HON. C. S. ROLLS IN FLIGHT.

A series of postcards was produced in Wolverhampton to celebrate the meeting, showing photographs of the various pilots and drawings of their aircraft superimposed on a picture of the racecourse. This is the Short-Wright biplane of Charles Rolls, who had recently formed a motor car company with one Henry Royce. Two weeks later, at a flying meeting at Bournemouth, his aircraft crashed and he was killed.

George Cockburn (Aviator's Certificate No.5) winning the 'Get-off' competition by taking off after a run of only 100 ft 5 in in his Farman biplane.

Lancelot Gibbs (French Aviator's Certificate No.82) flying his Farman biplane, which he crashed during the meeting while leading the duration competition with an aggregate 1 hr 13 mins 5 secs, which was then surpassed by Grahame-White.

Graham Gilmour (French Aviator's Certificate No.75) flying his Bleriot by the 'control tower' which was erected in the middle of the racecourse, On 28 June, Gilmour had become the first Scotsman to fly in Scotland.

After the 1910 meeting, a few other aircraft were based at Dunstall Park. This is the Mann & Overton's monoplane, powered by a 35 hp Anzani engine. It was built in Pimlico and tested at Dunstall Park for a while.

The Star monoplane was heavily rebuilt, as shown here, and flown at Dunstall Park by both Joe Lisle, the son of the owner of Star Engineering, and by its designer, Granville Bradshaw. The Star was later flown by Bradshaw at Brooklands.

The largest aircraft in the world, the Seddon *Mayfly* at Dunstall Park in September 1910, with the Mann & Overton's monoplane beyond to give it scale. It was built to the design of Lt John Seddon of the Royal Navy by Accles & Pollock at Oldbury, using 2,000 ft of their tubing fabricated into 20 ft diameter hoops to provide the structure. It was powered by two 65 hp NEC engines in a central nacelle. On the first taxiing tests on 7 November 1910, the starboard axle broke, probably due to side loads as much as the 2,600 lb gross weight of the machine.

A close up of the two NEC engines and fuel tanks with the pilot's steering wheel set between them. Constant problems with these engines meant no serious attempt was ever made to fly the *Mayfly* and by May 1911, Seddon had been recalled from paid leave granted by the Admiralty, and the aircraft was eventually broken up.

Sunbeam's company pilot, Jack Alcock, later the first pilot to fly the Atlantic non-stop, in front of the Farman biplane bought by Sunbeam to test their new V8 aero engine. The aircraft was usually based at Brooklands but Alcock did fly it to Dunstall Park, though the location of this picture is not known.

During the First World War Sunbeam built nearly 700 aircraft, including fifteen Short Bombers, powered by their own 240 hp V12 Ghurkha engine. This was the first one, serial 9356, shown at Dunstall Park with a full bomb load, and a non-standard fin and rudder.

Sunbeam also built over 500 Avro 504s, mostly with rotary engines, but they fitted this one with their own 100 hp water-cooled Dyak engine, and it is shown at Dunstall Park in 1919. Seven Dyak-powered 504s were operated very successfully in Australia, and one was the first aircraft used by Qantas.

Two

Bass's Meadow, Burton-on-Trent

In 1910 and 1913, flying meetings were held at Bass's Meadow, Burton-on-Trent. Forming part of an island made by two arms of the River Trent, the flying ground was 700 yards long by 300 yards wide. Messrs. Bass, Ratcliffe & Gretton were hosts, and provided their excellent refreshments for the flyers.

Six aircraft took part in the first meeting from 26 September to 1 October 1910. There were the Farman biplanes of Mlle Helene Dutrieu, Edouard Beaud and Brunneau de Laborie, the Goupy biplane of Emile Ladougne and the Bleriot monoplanes of Julien Mamet and Paul de Lesseps. The 1913 meeting, from 1 to 5 August, was organised by Frederick Handley Page and had only three aircraft; apart from Handley Page's own monoplane, to be flown by Ronald Whitehouse, there were the Avro 504 of Fred Raynham and the Bleriot monoplane of Sydney Pickles. On the Bank Holiday Monday about 10,000 people attended, the sixpenny enclosure being packed, though this did not compare with the 1910 meeting when 29,000 people attended on one day with people coming from all over the Midlands.

Bass's Meadow continued to be used intermittently for flying, especially by the post-war barnstormers. Alan Cobham's Flying Circus visited Burton on 3 August 1933, and on 19 September in both 1934 and 1935. In July 1929, the field featured a rare event - the first flight of a new aircraft. The prototype Civilian Coupe, built in Moor Street, Burton, made its first flight from there and made further test flights the following month. After the Second World War the former RAF airfield at Tatenhill was a more appropriate venue for local flyers.

The first flying meeting held in Staffordshire and the Farman biplane (No.1) of Edouard Beaud flying over Bass's Meadow, Burton, in September 1910. (*Burton Daily Mail*)

BRUNNEAU DE LABORIE AT BURTON. 69.

All the pilots at the meeting were French, in contrast to Dunstall Park's meeting earlier in the year, and this is Brunneau de Laborie, in front of his Farman biplane (No.2). (*Burton Daily Mail*)

Brunneau de Laborie crashed during the meeting after only a minute of flight. Taking off for the first time on Thursday 29 September he reached a height of 100 ft, when a wing dropped and he side-slipped into the ground. He clambered unhurt from the wreckage. (*Burton Daily Mail*)

The Bleriot monoplane (No.5) of Julien Mamet outside his hangar. Paul de Lesseps, the son of the builder of the Suez Canal, also brought a Bleriot to the meeting. Mamet made one flight to Lichfield and back, but when de Lesseps attempted to emulate him, he became lost and had to land at Grange Farm near Lichfield as darkness descended. The farmer let him store the aircraft in his barn and then charged locals an entry fee! On his return flight on Friday 30 September, de Lesseps became lost again, and landed at Nottingham! (*Burton Daily Mail*)

"Goupy Biplane"
BURTON AVIATION MEETING. 1910.

7.

The Goupy biplane (No.6) of Emile Ladougne. This aircraft was a trendsetter in that it had forward staggered wings, for better forward and downward visibility but, unusually, it also had a staggered biplane tail. (*Burton Daily Mail*)

The second flying meeting at Bass's Meadow was held over the August Bank Holiday week in 1913, and featured three aircraft including the 50 hp Avro biplane, precursor of the famous Avro 504, flown by Fred Raynham, who became one of the most famous test pilots in the country. (*Burton Daily Mail*)

Sydney Pickles, an Australian who came to Britain in February 1912 to learn to fly, brought his British-built Bleriot monoplane. Pickles was not a stranger to the county as, on 20 April, he flew at Dunstall Park and then flew up to Newcastle-under-Lyme for three days of flying. On 22 April he escorted the King and Queen on their journey from Crewe to Newcastle, swooping low over the royal procession for the last four miles. (*Burton Daily Mail*)

Sydney Pickles flying over the Burton skyline and in the photograph below, the aviators relaxing. From left to right: Mr Cates (Shell), Mr Lane, Fred Raynham, Mr Meredith, E.R. Whitehouse, Sydney Pickles, Mrs Whitehouse, Mr Murray, Mrs Goring and, standing behind, Frederick Handley Page.

Crowds around the Avro in which Raynham won the fastest take off competition in 6.25 secs and the race to Repton, but came second to Whitehouse in the six lap Round the Island race. (*Burton Daily Mail*)

Helpers hold back Sydney Pickles' Bleriot as he runs the engine. He won the highest flight competition with 6,100 ft and took up a number of passengers during the meeting. (*Burton Daily Mail*)

NATIONAL AVIATION DAY CRUSADE

SIR ALAN COBHAM'S
GREAT
AIR DISPLAY

20 Spectacular Events - Twice Daily

Including—

Demonstration of advanced aerobatics and upside-down flying.
Thrilling Parachute Descents.
Wireless-controlled flying and aeroplanes dancing in the air to music.
Miniature Schneider Trophy Race.
Daring delayed-drop parachute descent.
Demonstrations of the Autogiro.
Aerobatic thrills for passengers.
Smoke Stunting—The pilot's evolutions traced by smoke.
Thrilling exhibitions of wing-walking and trapeze acrobatics in the air.
An opportunity for a war-time pilot.
Humorous and surprise items, "The Battle of the Flours" and "Aerial Pig-sticking."
Speed and height judging competitions.
Grand Fly-past and Parade of Aircraft.

Cruises in Air Liners. Flying Lessons Short Flights in the "Moth" and other open aeroplanes. Autogiro Flights. Passengers carried in many display events.

Shobnall Road, Airport Site
BURTON-ON-TRENT
FOR ONE DAY ONLY!
THURSDAY, AUGUST 3rd, 1933
Two Complete Displays 2-15 and 7 p.m.
Admission 1/3. Children 6d. Cars 1/- Flights from 4/-.

"MAKE THE SKYWAYS BRITAIN'S HIGHWAYS"

The poster for Alan Cobham's 1933 visit to Burton. Air Displays are rather more expensive these days. (The Graham Nutt Collection)

Burton Daily Mail

THE PAPER WITH THE RECORD SALES.

READ IN EVERY LOCAL HOME.

No. 8863   REGISTERED AT THE GENERAL POST OFFICE FOR TRANSMISSION ABROAD.   FRIDAY, JULY 19 1929.   Telephones: BURTON Nos. 485, 486 and 487. SWADLINCOTE OFFICE, 705.   ONE PENNY.

POLICEMAN ON RUNNING-BOARD.

STORY OF MOTORIST'S DASH FOR LIBERTY.

BURTON DRUNKENNESS CHARGE DISMISSED.

Said to have ignored a policeman's advice not to drive a car, to have driven the car through the town with the policeman on the running board, and ultimately to have run away from the officer and been caught, Raymond John Shilton, of Middle Place, Newhall, answered a charge of being drunk in charge, at Burton Police Court, to-day.

Mr. Fisher Jesson prosecuted, and Mr. W. Goff, of Birmingham, defended.

Dr. J. B. Stanley said he examined the defendant at the station on Monday night. He put him through various tests which he thought were necessary. He was not in a fit condition to drive, as his judgment was impaired. His car was a Bentley.

MADE IN BURTON.

Mr. H. D. Boultbee (left) and Mr. A. P. Hunt, managing director and acting works manager respectively of the Civilian Aircraft Co., Burton, with the "Civilian" coupe, which has been made in Burton, to Mr. Boultbee's design. The company, at present in an experimental stage, is intended to meet the new private demand for aeroplanes. Mr. Boultbee is the son of the Rev. H. Travers Boultbee, a former vicar of Holy Trinity Church, Burton.

RAN OVER CYCLE.

VISITING MOTORIST WHO DI[D] NOT STOP.

Failing to obey a policeman's sign to stop at Bargate's corner, Ha[rry] Cockain, of Shakespeare Street, Loug[h]borough, collided with a cycle, and Burton Police Court to-day he paid fine of £1 and 15s. costs.

Harold Griffin, the High Str[eet] jeweller, said that on June 30 he w[as] cycling along High Street and was s[ig]nalled to stop by the policeman. did so, dismounting.

He heard a shout, turned round a[nd] saw a motorist coming behind him.

The car went on and ran over bicycle.

P.C. Carson said the motorist pass[ed] him for three-quarters of the ca[r] length despite the signal to stop.

In a statement, defendant (who [did] not appear), said the cyclist had su[d]denly turned to the right in front of car.

Front page news in 1929 and the prototype Civilian Coupe, G-AAIL, in the Moor Street premises of A.S. Briggs & Co., where it had been built, about to make its first flight from Bass's Meadow. It had been designed by Burton-born Harold Boultbee, who had been Handley Page Assistant Chief Designer from 1922 to 1928. (The Graham Nutt Collection)

The two-seat Coupe was powered by a 75 hp ABC Hornet engine for its first flight in July, and for further flights from Bass's Meadow the following month piloted by Herbert Sutcliffe, the chief instructor of the Midland Aero Club. The aircraft is seen here at Heston. The Coupe was put into production at Hull Airport, but only four more were built.

Three

Meir Airport, Stoke-on-Trent

There was a wave of municipal airports opened in the 1930s, many inspired by Alan Cobham and his National Aviation Day Displays. Stoke built the first in Staffordshire, opening on 18 May 1934 next to the Uttoxeter Road at Meir. It was a small grass airfield with a single hangar and clubhouse, occupied by the light aircraft of the North Staffordshire Flying Club.

For three years from 1935, Railway Air Services operated a 'request stop' on its Glasgow-Belfast-Croydon Route. The stationmaster at Meir Station had a short wave radio and if there were any waiting passengers, he would inform the pilot of the Dragon Rapide as it flew over Tunstall. He would then have to go to the airport to weigh them.

From 1938, No.28 Elementary and Reserve Flying Training School began to operate from Meir, using Tiger Moths and Hawker Harts and, from 1940, No.1 Flying Practice School also worked from Meir using Hectors and Hinds, but did little flying; the main wartime resident became No. 5 Elementary Flying Training School and its Miles Magisters.

Half a mile from Meir the Government built a Shadow Factory at Blythe Bridge which was administered by Rootes Bros. There was a taxiway to the airfield and a short runway was built from which to test fly the aircraft. Production started in December 1941 when Blenheim V production was transferred from Speke, and 718 Blenheims were completed, followed by 260 Bristol Beaufighters. In addition 2,178 Harvards and Mustangs, shipped from America in crates, were assembled and test flown at Blythe Bridge.

After the Second World War no new flying club took up residence at Meir except for the Staffordshire Gliding Club, which used a Tiger Moth tug, and No.45 (later No.632) ATC Gliding School. The airfield was used by Staffordshire Potteries' Dove for a while, but residential and industrial development all around saw it close in the mid 1970s.

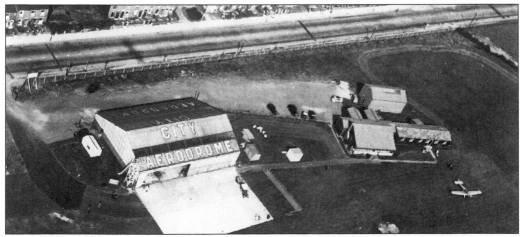

An aerial view of Meir Airport just after it opened in 1934, showing the single hangar and clubhouse and a Miles Hawk light aircraft of the North Staffordshire Flying Club. The road in the background is the A50.

A crowd of girls from Moorland Road Senior Girls School, Burslem, around a Miles Hawk at the airfield in September 1938. They travelled from Burslem by train and then walked from Meir Station.

A Taylorcraft monoplane, G-AFJP, of the North Staffordshire Aero Club, photographed in 1938.

SATURDAY, OCTOBER 12, 1935   (Registered for Tr
as a Newspa

:FUSES TO LEAVE

TO
R

or

ND

ND

ve been
ra and
mperor
Dedjas-
Gugsa

I

tl
s
S
I
R
L
p

L
w
of
L

w
ar
ac
ar

NORMAN WAINWRIGHT, the noted Hanley swimmer, and Mr. L. H. Koskie, his coach, leaving Stoke-on-Trent by air this morning for Belfast, where he will give a swimming exhibition to-night.

A Railway Air Services Dragon Rapide at Meir Airport picking up two passengers for Belfast, Norman Wainwright, the swimmer from Hanley, and his coach L.H. Koskie. On the right is Arthur Rogers, the LMS representative from Meir Station, who will have radioed the aircraft to land and then gone to the airport to check the tickets and weigh the passengers.

Stoke on Trent - Meir

Flugplatz

Lfl. Kdo. 3   Juli 1941

Länge (westl. Greenw.): 2° 06′   Nördl. Breite: 52° 58′ 30″
Zielhöhe über NN 183 m

Karte 1 : 100000

GB/E 16 b

Maßstab etwa: 1 : 11 000

500   0   500 m

A German bombing photograph of Meir Airport, taken in July 1941. The Rootes Bros. Shadow Factory can be seen under construction at the bottom, with the taxiway leading to Grindley Lane. The flight sheds were located under the small cloud and the RAF hangars are next to the Uttoxeter Road, which runs through the village of Blythe Bridge on the right.

28

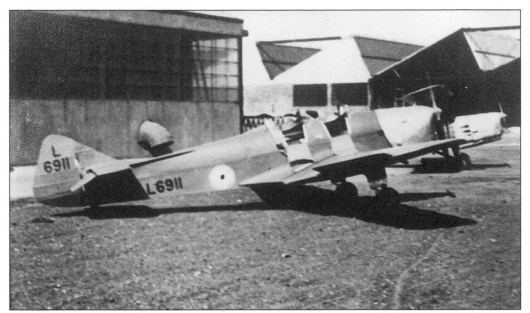

A Miles Magister, L6911, of No.5 Elementary Flying Training School, which was the main resident unit at Meir during the Second World War. This photograph was probably not taken there; the aircraft behind is a Blackburn B.2, which was normally only used by No.4 EFTS at Brough.

The two Bellman hangars alongside the Uttoxeter Road at Meir, probably just before the Second World War, as neither they nor the aircraft seem to be camouflaged. However, the presence of three Miles Magisters alongside the Hawker Hart/Hind trainers might indicate a later date. These hangars are now incorporated into the Staffordshire Potteries' complex.

One of hundreds of Bristol Blenheims built by the Rootes Bros. Shadow Factory at Blythe Bridge, on a test flight. This is a Blenheim Mk.V, DJ702, one of 718 built at Blythe Bridge.

A photograph which is out of chronological order in that this USAF Stinson, serial 417192, landed in the early 1950s. It was on its way north to Burtonwood, and on a murky day the pilot landed at Meir to ask the way.

A New Year dinner in the staff canteen at the Shadow Factory, probably in 1944, with the Rootes Bros. Home Guard unit and an American 8th Air Force Unit which was stationed further along Grindley Lane, probably associated with the Harvards and Mustangs assembled at Blythe Bridge. Fifth from the left on the top table is Flg Off Shultz, the CO of the Home Guard and also the chief test pilot for the factory. First on the right hand table is Dr Boyd (wearing glasses) and next to him is Mr Clarke, one of the men who came originally from Speke and later became a local optician.

A typical flying scene at Meir in the post-war years, featuring a group of Slingsby T.21 and T.31b gliders of No.632 ATC Gliding School with a visiting red Olympia glider, owned by Bill Nadin and Harry Primrose, and usually resident at Castle Bromwich.

This group of three posing in front of a T.31 are, from left to right: John Ash, Mike Ruttle and Frank Bott.

A T.31, WT869, just taking off on what was probably a first solo, and the last thing the student needed was a small black dog running alongside! The hangar was on the north side of the airfield with the clubhouse close by.

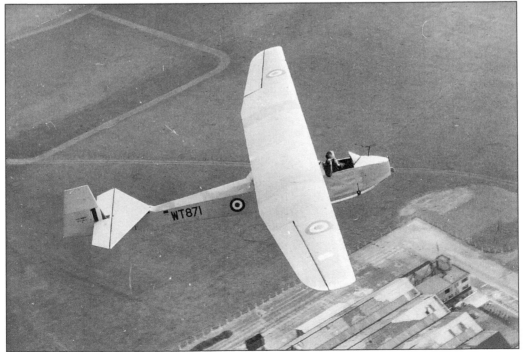

A T.31, WT871, being photographed from a T.21b, flying in formation over Staffordshire Potteries. Note that Mike Ruttle is in the front cockpit of the T.31 taking the reverse photograph, with Charles Webb piloting behind him. This was taken on 7 June 1953.

A T.21b flying over the Uttoxeter Road, to the south of Staffordshire Potteries, to land behind a T.31.

A row of ATC cadets by the ATC hut in August 1957, with their instructors kneeling in front. Front row, left to right: Bill Beeston, Charles Webb, 'Hutch' Hutchinson, Roger Smith, John Davies, Mike Ruttle.

The Auster owned by Sqn Ldr William Nadin, the District Gliding Officer, who used it to visit other Midland Schools, such as those at Fradley and Castle Bromwich. Phil Howorth is swinging the prop on a cold morning.

The Staffordshire Potteries' de Havilland Dove, G-ALCU, outside their complex. This aircraft was sold in 1960.

Tiger Moth G-AHUE, which towed the gliders of the Staffordshire Gliding Club aloft, at Meir in the winter of 1969/70. The pilot was possibly Ken Sherriff, who lost his life in the same aircraft the following year when it collided with a glider. The houses in the distance are on Sandon Road, Meir Heath. (Eric Ralphs)

An aerial view of the Staffordshire Potteries' complex at Meir in 1979. If this is compared with the first photograph in this section, the original hangar can still be seen next to the gate.

Four

Walsall Airport

Walsall commissioned its own airport in the 1930s, with a site chosen by Alan Cobham at Bosty Lane, Aldridge. Before this, flying was usually undertaken by visiting barnstormers from Calderfields Farm. Work on the new municipal airport began on 1 June 1933, and a grass airfield of about 220 acres was created with a small hangar and clubhouse on the northern side. The first resident was the Walsall Aero Club, which used a single Miles Hawk trainer, but after getting into financial difficulties the South Staffordshire Aero Club was created to take things over, and was notable for having the only female flying instructor in the country, Mrs Gabrielle Patterson.

Helliwells Ltd moved into a factory unit on the west side of the airfield and during the Second World War, as well as producing sub-assemblies for various manufacturers, they were the 'sister' company for the Douglas Boston, undertaking local modifications. Number 43 Gliding School occupied the airfield, providing air experience flights to ATC cadets in Kirby Kites and Kadets.

After the Second World War Helliwells were engaged in overhauling Seafires and Harvards, and took over the running of the airfield as there was no resident flying club. Number 43 Gliding School moved to RAF Lichfield in 1947. When the last refurbished Harvard flew away in 1956, Helliwells moved their operation to Elmdon, and the airfield closed for flying.

Avro 504K, G-EASF, at Calderfields Farm in the 1920s, where it was giving joy flights, as they were called. Such companies invariably used Calderfields Farm before Walsall Airport was opened at Aldridge.

The Desoutter cabin monoplane, G-ABMW, which was based at Walsall Airport in 1934. Behind is one of the South Staffordshire Aero Club's Miles Hawks.

Members of the South Staffordshire Aero Club in front of their two Avro Avians, G-AAVM and G-ABCD. The lady is Mrs Gabrielle Patterson, the flying instructor, with the chairman of the club, Norman Parkes, standing behind her.

Mrs Gabrielle Patterson (left) serving in the Air Transport Auxiliary in 1940, along with Mrs Grace Brown. The location of this photograph is not known, but the ATA will have delivered Tiger Moths to several Staffordshire airfields during the Second World War.

The two-seat Tipsy Belfair monoplane, OO-DOP, in the Walsall club hangar just after the Second World War. A Belfair had been demonstrated at Walsall in 1939, but there were no buyers.

An aerial view of Walsall Airport taken in 1948. The hangars on the left were operated by Helliwells Ltd, which undertook modifications to Douglas Bostons during the Second World War, and Seafires and Harvards after it ended. The footpath from Helliwells to the clubhouse and hangar can be seen crossing a runway. The company's experimental hangar is just above the copse to the right.

Taylorcraft, G-AGZN, outside the civil hangar on 13 April 1947, on the day when the Midland Model Aircraft Rally was held at Walsall.

Engine running, Tiger Moth G-AHLA awaits its pilot on the grass airfield on a sunny but hazy morning. This picture epitomises British club flying in the years immediately after the Second World War.

Percival Proctor G-AHGT, which was a visitor to Walsall that morning.

An Avro Anson, NK328, fitted with assorted experimental aerials. This aircraft was assigned to Helliwells in connection with work with the Blind Landing Experimental Unit.

Five

Wolverhampton Airport, Pendeford

The last of the three Staffordshire towns to build an airport was Wolverhampton, choosing a site at Pendeford recommended by Alan Cobham, despite the fact that he usually used Perton for his flying displays. It was another grass field, with a single hangar and a small clubhouse/terminal, used intially by the Midland Aero Club, operating two sites with Castle Bromwich.

Boulton Paul Aircraft from Norwich chose Pendeford for their new factory and construction began in 1935, the move being completed by the end of 1936. The company built 106 Hawker Demon fighters, 136 Blackburn Rocs and then 1,062 of its own famous Defiant fighters. The Defiant was followed by 692 Fairey Barracudas, but many other aircraft were seen at Pendeford having their Boulton Paul gun turrets fitted. After 1945, 270 Wellington bombers were overhauled and converted to trainers, and then the Balliol went into production.

During the Second World War No.28 EFTS operated from Wolverhampton with up to 108 Tiger Moths on strength; four new hangars and many other buildings were constructed. After 1945 Wolverhampton Aero Club was formed and operated alongside No.25 Reserve Flying School until that was closed in 1953.

Don Everall Aviation took over the running of the airport and even operated scheduled airline services for a short while, as did Derby Aviation. The encroaching houses threatened the airport's existence and when a Dove on hire to Dowty crashed into a house in 1970, the writing was on the wall. Soon after the airport was closed, and Pendeford is now just another large housing estate.

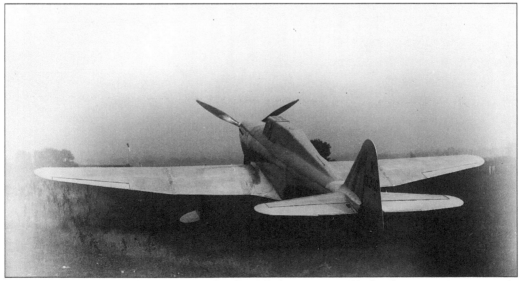

The prototype Boulton Paul Defiant, K8310, stands waiting on Wolverhampton Airport ready for its first flight in August 1937, its unpainted surfaces shining in the hazy sunlight. The first flight on 11 August ended prematurely when an oil pipe in the propeller hub burst, and the fuselage was covered with oil.

The official opening of Wolverhampton Airport on 25 June 1938. The aircraft ranged in front of the clubhouse and hangar are Moth Major, G-ACOG, Miles Falcon Six, G-ADLC and Avro Tutor, G-ACFW.

Line up of dignitaries on the opening day. From left to right: the mayor, R.E. Probert, Mrs Probert, Mrs Kidson, Mr B. Kidson (Chairman of the Airport Committee) Mrs A.E. Clouston and Flg Off A.E. Clouston who performed the official opening, having just set a record flight to South Africa and back.

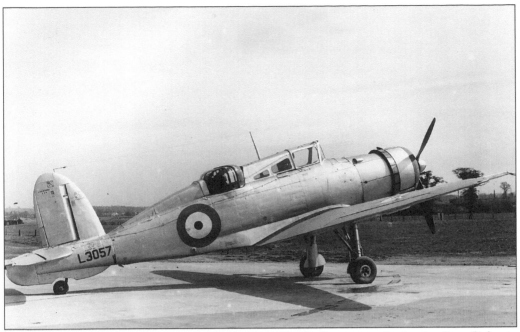

The prototype Blackburn Roc, outside Boulton Paul's factory, ready for its first flight in December 1938. Boulton Paul built all 136 Rocs, following the production of 106 Hawker Demons built at Pendeford.

Defiants awaiting delivery outside the camouflaged flight shed at Boulton Paul, in May 1940. The one on the right, L7005, was one of the twelve of No.264 Squadron, which shot down thirty-seven German aircraft over Dunkirk on 29 May 1940, and was itself forced down in Kent during the Battle of Britain.

An aerial view of what had become RAF Wolverhampton in 1941. Four T2 hangars and a large camp had been built to house No.28 Elementary Flying Training School and its Tiger Moths, twenty-seven of which can be counted on the ground.

Sir Stafford Cripps (the tall man in the centre) on a visit to Boulton Paul in 1943. On his right is Ralph Beesley, general manager, and second on his left is J.D. North, managing director. Many of the workforce were women by this time.

The Fairey Barracuda under production at Boulton Paul in May 1942. The Barracuda replaced the Defiant on the production lines; 692 were built during the Second World War.

A dummy Staffordshire airfield, next to the dummy Boulton Paul factory built a mile north of the real thing, complete with dummy aircraft and even dummy cars on the car park. It did not fool the Germans who showed both the dummy and the real thing on their bombing maps.

After the Second World War Boulton Paul refurbished 270 Wellington bombers and converted them to navigation trainers. This is one in the Rigging Shop in 1949 with Esme Johns (later Mrs Slim Bunkell), a cover and dope operator. Esme was a cousin of Glynis Johns, the actress.

Corporal Ray Simpson of No.2078 Squadron ATC 'C' Flight (Wednesfield) receiving help with his parachute at Pendeford in 1947, just before an air experience flight.

The prototype Boulton Paul Balliol, VL892, which first flew with a Bristol Mercury radial engine, shown flying over Pendeford after being fitted with its Armstrong-Siddeley Mamba turboprop in 1948. The airfield is bottom right.

The first pre-production Balliol T.2, VR590, with the Merlin engine, outside the factory in 1949. On the right is the mock-up of the P.119 jet trainer project.

The Goodyear GA-2 Duck, NC5506M, visiting Pendeford in June 1948. One of nineteen of these 2/3-seaters built by Goodyear in America for experimental purposes, it was powered by a single 145 hp Franklin engine.

Miles Aerovan, G-AJKP, being refuelled at Pendeford, with the control tower in the background.

Avro Anson, VV296, of the No.25 Reserve Flying School, based at Wolverhampton after the Second World War.

The spacious production lines of the last aircraft built by Boulton Paul, the Balliol trainer. The third production aircraft, WF991, is in the background on the left. A total of 199 Balliols of all Marks were built at Wolverhampton, plus another thirty by Blackburn Aircraft at Brough.

One of the few jets to operate from Pendeford's grass runways, the Nene-powered Vampire, TG270, on 5 January 1950. Boulton Paul had modified the tail and the intakes. Note that the control tower in the background is just being built to replace the single storey watch office alongside.

The Miles Hawk Trainer, G-AHNV, of the Wolverhampton Aero Club, shown in 1951. After the Second World War the airport was operated initially by Wolverhampton Aviation, who were Miles agents.

Spitfire G-AISU, at Pendeford in 1950, to take part in the King's Cup Air Race, flown by Miss I.M. Sharpe.

The Mosscraft MA.1, G-AEST, flown in the race by its designer, W.A. Moss. Unfortunately he crashed while rounding the turn at Pave Lane, Newport, and was killed.

Boulton Paul P.111, VT935, on the taxiway from the factory to the airfield in February 1950. This experimental delta wing jet undertook taxiing trials on Pendeford but did not fly from there.

Percival Proctor 3, G-AMAN, at Pendeford for the Goodyear Trophy Air Race in May 1952.

The rare Miles Nighthawk, G-AGWT, entered for the Goodyear Trophy Air Race in May 1951.

The beautiful Percival Q.6, G-AHOM, named *Ductility*, was a resident of Wolverhampton Airport for some time, and was the corporate aircraft of Ductile Steels Ltd.

Percival Prentice, VS698, of No.25 Reserve Flying School. The Prentice had been intended as the basic element of the RAF's flying training programme.

The advanced element of the RAF's flying training programme following the Prentice was intended to be the Balliol T.2. This is the first production aircraft, WF989, taxiing out at Pendeford in May 1952. Note Ductile Steels' Q.6 in the background.

An unusual visitor to the 1952 Air Races was this Fieseler Fi.156 Storch, VP546: a captured German aircraft used by the RAE.

An Avro Tutor, G-AHSA, at the 1952 races, just like the one which attended the airport's open day fourteen years before.

The BEA Sikorsky S-61, G-AJHW, which had brought the Minister Of Aviation, a certain John Profumo, to the 1952 Air Races. This aircraft would be standing on the Droveway these days.

In the mid '50s both Don Everall Aviation, who had taken over the running of the airport, and Derby Aviation, operated scheduled services for a short while to the Channel Islands and the Isle of Man, using Dragon Rapides. This is one of the Don Everall Rapides, G-AGDP, standing awaiting passengers.

Austers were a familiar part of the Pendeford scene in years following the Second World War. This is Autocrat, G-AGVP, at the 1952 Air Races.

Miles Gemini 3A, G-ALCS, on a lovely sunny day at the 1952 races. This was the first of six Geminis assembled at Pendeford by Wolverhampton Aviation, in April 1950, from parts acquired after the collapse of Miles Aircraft. It had the construction number WAL/c.1001, and in 1957 was converted into a Gemini 3C.

Heavy metal taking part in an RAFA Air Display during the 1960s. Four F.111s of the USAF formate on a KC.135 Stratotanker.

The largest aircraft ever to land at Pendeford, a Blackburn Beverly at the same RAFA Air Display. Note Pendeford High School just behind its tail.

A civil Percival Provost, G-ASMC, outside the single Bellman hangar which was normally used for aircraft, with a Luton Minor in the background.

The most common aircraft seen at Pendeford during its thirty-five years of existence, a Tiger Moth, G-AHVV, owned by M.J. Lawrence outside the T2 hangar in the '60s. This aircraft is still flying.

The airport manager, Eric Holden, in front of Don Everall's new Piper Cherokee G-ARVR, in 1961, with one of their Piper Colts, G-ARON, alongside.

The Boulton Paul Association's Balliol Project, the restoration of the last RAF T.2 in a building at the factory where it was made, now part of Dowty Aerospace, Wolverhampton, and one of the centre-pieces in their planned Staffordshire Aviation Museum.

Six

RAF Halfpenny Green

The expansion of the RAF lead to the construction of a number of new airfields in Staffordshire and one of the first of these was Bobbington, later renamed Halfpenny Green because of confusion with Bovingdon. Construction was started in 1939, with three tarmac runways and seven Bellman hangars, and the station opened in February 1941 as the home of No.3 Air Observers Navigation School. Initial complement from May was fifty Blackburn Bothas, but these began to be replaced by sixty-six Avro Ansons from August.

In 1942 the unit was renamed the No.3 (Observer) Advanced Flying Unit, and was joined by the School of Flying Control from RAF Watchfield. Flying ceased in November 1945 and the unit was closed the following month. The Korean War caused the airfield to be reopened, with the runways resurfaced and the return of the Avro Anson with an Air Navigation School, but it closed again in 1953.

In 1960 Ted Gibson, former Chief Instructor with No.25 RFS at Wolverhampton, inspired the reopening of the airfield as a general aviation centre. Fighting red tape and a certain amount of local opposition, Halfpenny Green has become a thriving general aviation centre, now operated by Bobbington Estates Ltd, with four flying schools, a helicopter school, the Police Flying Unit and the Virgin Airships hangar. It is easily the busiest airfield in the county, and because of its location at the southern tip of Staffordshire it also serves Shropshire, Worcestershire and the West Midlands, as the major general aviation airfield in the area.

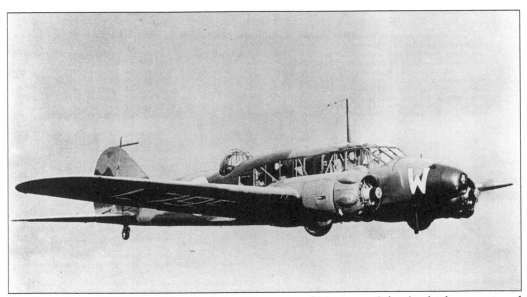

An Avro Anson I, L7951, of the No.3 Air Observer and Navigator School, which re-equipped with Ansons at Halfpenny Green from July 1941. The Anson was the most widely used aircraft at Halfpenny Green throughout the Second World War.

Blackburn
AIRCRAFT

Initial equipment of the No.3 Air Observer and Navigator School at Halfpenny Green was the underpowered and unloved Blackburn Botha, as shown in this contemporary Blackburn advertisement.

This Anson, N9908, is shown flying with No.48 Squadron in its maritime reconnaissance days. It later served at Halfpenny Green with what had become No.3 (Observer) Advanced Flying Unit.

Many years later, in January 1969, Tippers Air Transport bought eight ex-RAF Anson C.19s, and parked them at Halfpenny Green while looking for a buyer, giving the airfield something of its wartime flavour. Most had been scrapped by 1969 but two survive: VL348 (G-AVVO) at Newark Air Museum and the nose of VP519 (G-AVVR) stored by The Aeroplane Collection in Manchester.

A plan of the airfield showing that there were originally seven Bellman hangars, which were reduced to three when the airfield was reopened as a civil airport, to avoid paying rates.

A long-term resident at Halfpenny Green was this Percival Proctor, G-AOBI, which belonged to the Air Scouts.

The Air Scouts also acquired this Avro Anson C.19, VM325, which was their pride and joy for many years. After it was attacked by vandals it was donated to the Midland Air Museum at Coventry where it is in store awaiting restoration.

A vintage scene by the control tower in the mid '60s, with a visiting Fairchild Argus and behind it, the Piper J.3C Cub, G-ASPS.

Also visiting in the same vintage fly-in were this Currie Wot, G-APWT, and behind, the Miles M.11A Whitney Straight G-AFGK; in the background is the Miles M.3A Falcon, G-AEEG.

The Goodyear Trophy Air Race was revived at Halfpenny Green during the 1960s, and this is the 1968 appearance of Charles Masefield's American-registered Mustang, in the days when Mustangs were very rare beasts indeed.

Formula One short-circuit air racing was also introduced to this country and these are two Cassutts in the 1970 race, G-AXEA (No.40) was all-red and G-AXDZ (No.41) was all-orange.

INTERNATIONAL
AIR RALLY - AIR DISPLAY
& GOODYEAR AIR RACES
1970

AT HALFPENNY GREEN AERODROME
SUNDAY 30th AUGUST—MONDAY 31st AUGUST

HALFPENNY GREEN INTERNATIONAL AIR RALLY

—

**THE GOODYEAR INTERNATIONAL
HANDICAP AIR TROPHY RACE**

—

**THE GOODYEAR INTERNATIONAL
FORMULA AIR RACE**

—

THE 99's LADIES HANDICAP AIR RACE

2|6d
OFFICIAL PROGRAMME

The cover of the 1970 programme featuring two Rollason Beta Formula One racers. The following year the programme design was the same but the price was 20p.

The Piper Comanche, G-ATOY, *Miss Sunpip*, owned by Miss Sheila Scott, the aircraft in which she flew solo round the world. She took part in the Ninety Nines' ladies air race at Halfpenny Green in 1970, and won.

In the 1969 race this Cosmic Wind, G-ARUL, took off in the Goodyear Trophy Race, rounded the first pylon just beyond the runway and crashed after a high speed stall. Bill Innes, the pilot, received chest injuries and a broken leg.

This Czech-built Meta-Sokol, G-APUE, was a non-flying ornament at Halfpenny Green for some time.

The South Staffordshire Sky Diving Club was resident at Halfpenny Green for many years and one of the aircraft they frequently jumped out of was this Dragon Rapide G-AHJA.

Almost a wartime scene. The Battle of Britain Memorial Flight Hurricane, PZ865, stands ready for its slot in the flying display, the pilot's helmet draped over the windscreen.

The little Arrow Active single seat biplane, G-ABVE, a visitor to Halfpenny Green. Some of the Avro Ansons of Tippers Air Transport (TAT Ltd.) form a backdrop.

An aerial view of the crowd during the 1968 display. The Air Scouts' Proctor and Anson are in the foreground by the marquee, erected by what was then the Midland Aircraft Preservation Society, to display some of its relics. MAPS has now become the Midland Air Museum at Coventry Airport, one of the finest volunteer-run aircraft museums in the country.

Two Virgins on Halfpenny Green Airport

Proof that virgins fly from Halfpenny Green - in this case a hot-air balloon and airship of the Virgin Airship and Balloon Co. of Telford! Halfpenny Green has housed an airship hangar for a number of years.

The American A-60-II Blimp, N2022B, at the mast at Halfpenny Green in 1994. Compared with the Willows Airship, which flew from Dunstall Park over eighty years before, it is the same size but inflated with helium instead of hydrogen, and has far more efficient engines.

Keith Sedgewick's homebuilt 'Super Wot', G-AVEY, being erected at Halfpenny Green and having its engine run. Its first flight was from Halfpenny Green on 10 January 1971.

A Buccaneer approaching Halfpenny Green as if to land, with wheels and arrester hook down. It was only making a slow pass during the 1971 Air Show.

Seven

RAF Hixon

Hixon, to the east of Stafford, was built during 1941-2 as the home for No.30 OTU, equipped with Wellingtons; the first two arrived on 15 May 1942. As a night bomber training unit, Hixon frequently sent crews on missions to France and even Germany.

Seighford, on the other side of Stafford, was built as a satellite to Hixon and flying continued from both. Also based at Hixon was No.1686 Bomber Affiliation Flight, equipped initially with Curtiss Tomahawks with the shark-teeth noses which always seemed a feature of these aircraft. The BAT flight later received Hurricanes, Masters and Martinets.

Number 30 OTU moved out on 2 February 1945, and thirty-seven Beauforts of No.12 (P) AFU arrived from Spitalgate. However, this unit was closed in June, and Hixon became a satellite of 16 MU at RAF Stafford. It remained a storage site until 5 November 1957 when it closed and the airfield was sold off, being turned over to agriculture with the buildings becoming the basis for a new industrial estate.

A famous picture of thirteen No.30 OTU Wellingtons lined up at Hixon. The third aircraft bears the code 'KO' making it an ex-115 Squadron Wellington. The first aircraft, BK347, crashed at Whernside, Yorkshire, on 21 April 1944.

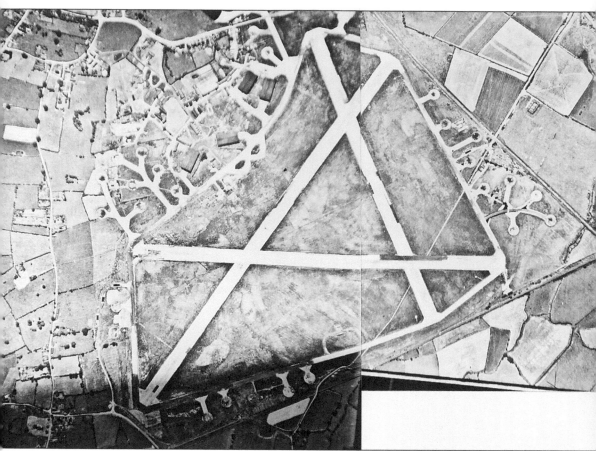

An aerial view of Hixon showing the typical three-runway layout and the curved taxiway in front of the hangars, where the photograph on the previous page was taken.

The ground crew for the Curtiss Tomahawks of No.1686 Flight, gathered around one of their charges during the winter of 1943/4.

A close up of six of the multi-national ground crew. From left to right: Flt Sgt Art Pincome (Canadian), Plt Off G.B. Chapman (British), Flt Sgt Jock Gleece (Scottish), Flt Lt L.M. Rolph (New Zealander), Flt Sgt K.Watts (Australian), Flt Sgt Val Turner (Australian).

One of the Tomahawks taking off from a very wintry Hixon. The BAT Flight provided 'targets' to train the bomber crews in defensive action when attacked by fighters.

Ground crew outside their hut at Hixon in 1943. From left to right, standing: Flt Sgt Val Turner, Flt Sgt Kenny Watts, Sgt Les Freeman and Flt Sgt Jock Gleece. Kneeling, Sgt Bill Harle and Flt Sgt Art Pincome.

The Stone ATC in 1944, including 'flights' from all the local villages, Hixon ATC being 'C' Flight 1436 Squadron. The CO, Flt Lt Cooksey, is seated centre (with pilots wings), to his left is Plt Off Beardmore, who was also the headmaster of Alleyne Grammar School, Stone. To his right is Plt Off Barker-Smith and then Plt Off Ralphs. ATC cadets from all over Staffordshire were given flights in No.30 OTU Wellingtons during the Second World War.

Sgt Les Freeman (Fritz) and Plt Off
G.B.'Chappie' Chapman outside a
Hixon nissen hut.

Three Tomahawks at dispersal with 'Chappie' Chapman on the nose, Jock Gleece on the wing
and standing, from left to right: Art Pincome, Kenny Watts, L.M.Rolph and Val Turner.

Eight

RAF Lichfield

RAF Lichfield, which was constructed in the years 1939-40, was the largest, and for some time the most important, airfield in Staffordshire. Built to the north of the city it is more usually known locally as Fradley. The station opened for flying on 23 April 1941 and the Wellington bomber was the most important aircraft in its history, serving No.27 Operational Training Unit during the Second World War and training mostly Royal Australian Air Force bomber crews. They went on many operational sorties, including the three famous 1,000 bomber raids. RAF Tatenhill near Burton-on-Trent was constructed as a satellite airfield, but proved unsuitable for Wellingtons, so in 1942, No.27 OTU's satellite became Church Broughton in Derbyshire. Eventually 27 OTU closed in 1945, but the Wellingtons returned in 1952 when No.6 ANS moved in.

Lichfield was also a large storage base and the first resident unit was 51 MU, being joined by 82 MU in 1941. The latter disbanded in October 1945, but at the end of the Second World War Lichfield had become one gigantic aircraft park, with 781 aircraft held in storage. This number increased by hundreds and by the end of the year there were aircraft overflowing into the surrounding fields, including 900 Typhoons and 500 Liberators.

Lichfield remained a storage unit, the only regular flying being the gliders of the ATC, until the Cold War saw its refurbishment and reopening as a training base with the arrival of No.6 ANS. Then 99 MU moved in for three years, but the station finally closed on 1 March 1957.

Flying reappeared when Ultraflight Aviation opened a microlight flying club, but this too was forced to move in 1994 and took up residence in a field on the other side of the A38, at Roddige.

A trainee bomber crew in front of their Wellington, 'R' for Robert, at No.27 OTU, Lichfield. On 30 May 1942, twenty-one such crews took off from Lichfield on the first 1,000 bomber raid against Cologne.

No.27 Operational Training Unit ground crew servicing a Wellington at Lichfield in 1942.

No.27 OTU ground crew fitters outside their billets. Most of these worked in the maintenance hangar under Flt Sgt 'Chippy' Carpenter. The only known name is Wally 'Lofty' Webb, third on the left, back row.

A happy bunch riding the port engine of a Wellington in 1943. Corporal Jenks is astride the spinner, Tom Bolton is holding the dog (name unknown), with Paddy Blair behind him, and Bert Hartley standing. 'Taffy' Thomas is in the glasses but the other two are unknown.

Flying Officer Kevin Cranley of the Royal Australian Air Force beside his Wellington in November 1944. Lichfield was particularly associated with Australians. There were 690 based there in September 1943 alone, and twenty-four lie buried in Fradley churchyard.

The Sergeants Mess at Lichfield in 1944, the reason for the dinner being unknown.

Another No.27 OTU RAAF crew in 1944. Left to right: Flt Sgt J. Reuter (wireless operator), Flg Off E. Minns (air gunner), Flg Off J. Harn (bomb-aimer), Flg Off K.Cranley (pilot), Flg Off G. Thomson (navigator), Flg Off E. Peace (air gunner).

The pilots and observers of No.22 Course in 1944. Their names are on the picture, but 'Turnhill' should read Turnbull and 'Raymond' should read Tarlton-Rayment.

Number 51 MU ground crew in front of Wellington 'F' for Freddie, in November 1944. Note that the front turret of the Wellington has been removed and replaced with a fairing, more commonly regarded as a post-war modification.

An aerial photograph of RAF Lichfield taken on 27 March 1948 from an altitude of 16,600 ft, showing the typical three-runway layout.

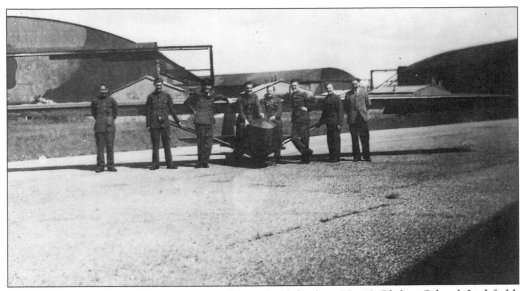

Members of No.2078 Squadron ATC in front of a glider from No.43 Gliding School, Lichfield, in June 1948. Third from the right is Sgt Ray Simpson.

Number 51 MU personnel outside J4 hangar, in front of one of their charges, an Avro Lincoln bomber. They were known as both 'Briden's Brit Bashers' and 'The 278 Terrors'.

A Valetta T.Mk.3 Navigation trainer taking off from RAF Lichfield in 1953. The Valetta was just beginning to replace the old Wellingtons. That was the same year 'Ben' Gunn flew the last Wellington T.10 to be delivered to the RAF, after conversion by Boulton Paul Aircraft.

A Gloster Meteor T.7, WH243, in one of Lichfield's hangars in 1953 after a landing accident.

A group from No.632 ATC Gliding School at Meir, on a course at Lichfield in 1953, alongside a T.31 glider. Mike Ruttle is the tall instructor by the cockpit.

A group of ATC gliders in a Lichfield hangar, including WE991 and WB960, with a Lincoln heavy bomber lurking in the darkness behind.

A very unusual resident at Lichfield during the 1950s was this civil de Havilland Gipsy Moth.

An RAF Auster towing a T.21b glider, WB960, from Lichfield back to Meir on 7 April 1953, after an ATC course. They are just crossing Blithfield Reservoir near Uttoxeter.

The cover of the programme for the 1955 Battle of Britain Day Air Display at RAF Lichfield.

Wellington T.10, MF627, of No.6 ANS, Lichfield, after crashing at Ughill, just north of Sheffield, on 17 October 1952. The two trainee navigators aboard were convinced they were on final approach to Lichfield and the pilot had the undercarriage and flaps down ready for landing. Suddenly the cloud became distinctly solid and they just had time to ram on the power and pull up the nose; the aircraft crashed in an upward slide. The crew were unhurt.

Nine

RAF Perton

There was an airfield at Perton, just north of Wolverhampton during the First World War. A relief landing ground was established on the Fern Fields, alongside Perton Ridge, from which No.38 (Home Defence) Squadron could operate its BE.2c, and later FE.2B, aircraft against Zeppelin attacks on the Black Country.

This same field was usually used by Alan Cobham when he brought his National Aviation Day Display to Wolverhampton during the 1930s, and in fact was the usual field for any flying that took place in the Wolverhampton area, so it was quite surprising when Alan Cobham recommended Barnhurst Farm for the town's new airport.

A new RAF airfield was built during the early days of the Second World War, to the west of Fern Fields, next to Yew Tree Lane. It was intended to be a fighter station but was never used for this purpose, and there was no resident flying unit on its opening on 21 August 1941. The Princess Irene Brigade of the Dutch Army moved into the accommodation units in the autumn, and in January 1942, the airfield became a satellite of RAF Shawbury and the Airspeed Oxford trainers of No.11 Service Flying Training School.

These were replaced by the Miles Masters of No.5 (Pilots) Advanced Flying Unit from Tern Hill in June 1942, but No.11 SFTS returned in September with a flight of fourteen Oxfords. This unit was split into two in 1943, and Perton became a satellite of No.21 (Pilots) Advanced Flying Unit, whose base was RAF Wheaton Aston.

Both Boulton Paul Aircraft and Helliwells of Walsall also used Perton, and its three tarmac runways, for flying Douglas Havoc bombers for which they undertook modifications

Perton closed in 1947 and is now a large housing estate.

The RAF Perton Maintenance Team in August 1944 alongside the solitary T2 hangar which was on the eastern side of the airfield. Only major inspections were done back at the parent airfields - Shawbury, Tern Hill, and then Wheaton Aston.

The aircraft most associated with Perton - the Airspeed Oxford trainer. This one is actually V3388, which was used post-war by Boulton Paul Aircraft as their corporate aircraft, and is shown at Staverton after going to the Skyfame Museum and reverting to military markings.

The instrument panel of a Miles Master I. Miles Masters of No.5 (P) AFU used Perton for a few months in 1942 as a satellite of RAF Tern Hill. The 'Basic Six' flying instruments in the centre of the panel were common to all RAF aircraft.

A contemporary advertisment for the Miles Master, showing the Mercury-engined Mk.II version, as it says at the bottom 'somewhere in England', which might even have been at Perton.

A Douglas Boston, BZ435, at Perton on 15 June 1944, having various items of equipment removed before being modified by Boulton Paul Aircraft. Pendeford was a grass airfield and the Bostons were better suited to Perton's tarmac runways.

Queen Wilhelmina of the Netherlands inspecting the Princess Irene Brigade of the Dutch Army at RAF Perton, where they were quartered in the accommodation units.

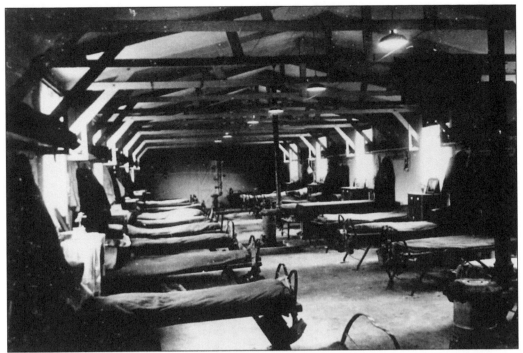

One of the Dutch Army's barrack blocks at RAF Perton. After the Second World War the accommodation was used to house Polish and other refugees, for a considerable time.

The headquarters building for RAF Perton, which was used after the Second World War as a private house, as shown here, but was later incorporated into a residential home.

An aerial view of RAF Perton taken in the 1960s. The control tower is the white dot just on the right hand side of the airfield. The road across the top of the picture is the Pattingham Road, running along Perton Ridge and the Fern Fields, the site of the First World War airfield, the field used by Alan Cobham's flying circus and others is the largest one next to this, near the right hand side.

Ten

RAF Seighford

Seighford opened in January 1943 as a satellite for No.30 OTU at Hixon, to replace the original intended satellite, Wheaton Aston, and had three tarmac runways and two T2 hangars. The Wellingtons stayed for less than a year however, because in October 1943 Seighford was handed over to No.23 Heavy Glider Conversion Unit at Peplow, so that its Horsa gliders, towed off by Albermarles, could have a site for practice landings. This lasted until January 1944 when the station became a satellite for No.21 (Pilots) Advanced Flying Unit at Wheaton Aston, with its Airspeed Oxfords.

Flying ceased in December 1946 and Seighford lay forgotten until 1956 when Boulton Paul Aircraft at Wolverhampton were looking for a new Flight Test Centre for their Canberra conversions, having been kicked out of Defford. They extended and resurfaced the main runway and erected a B1 hangar between the original T2s. For the next nine years all manner of Canberra and then Lightning modifications were to be seen at Seighford, as well as the Tay-Viscount fitted with the world's first fly-by-wire system.

With the cancellation of TSR.2, BAC took back all the Canberra and Lightning work for its own divisions and Seighford closed. However, a number of light aircraft have always been kept at Seighford and it has recently become the new home of the Staffordshire Gliding Club.

A Canberra B (I) 8 bomber on Seighford airfield in January 1960. The Canberra was a feature of Seighford for nine years from 1956 to 1965, when Boulton Paul Aircraft were the prime contractor for major Canberra modifications. This one, WT329, was modified as the prototype for the Royal New Zealand Air Force B.12.

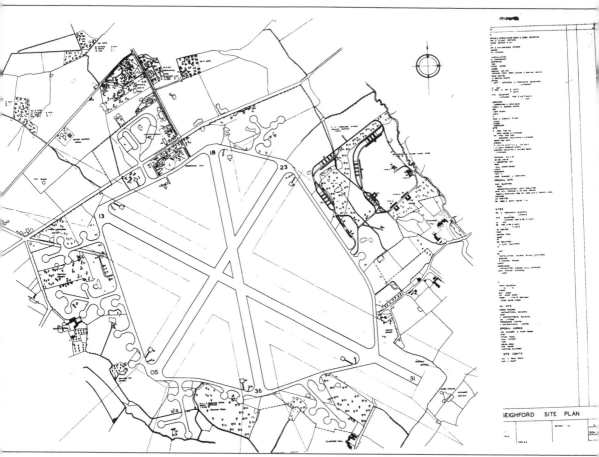

A wartime plan of Seighford airfield showing the three runways and the two T2 hangars to the north-west, on the other side of the lane between Eccleshall and Seighford.

On 28 December 1945, Flg Off Ted Croker (shown here) took off from Seighford in Oxford HN594 on a map reading exercise with Flg Off John Dowthwaite and Warrant Officer George Robinson. They crashed on Brown Knoll in the Peak District.

The remains of their Oxford on the snow-swept moor shortly after the accident. All three survived the crash, but only Ted Croker was capable of going for rescue. He struggled for a mile over the moor, mostly on hands and knees because of his sprained ankles, and eventually reached a farmhouse. The other two crew members were then rescued. In later years Ted Croker became Secretary of the Football Association.

A B1 hangar being constructed across an old taxiway between the original T2 hangars. Boulton Paul Aircraft were increasing the accommodation and resurfacing the runway to turn Seighford into their new Flight Test Centre.

Canberra B.(I) 8, WT327, at Seighford in October 1956, having had the Ferranti A.I.23 radar for the Lightning installed by Boulton Paul.

The 'stealth' Canberra, WX161, on Seighford's runway in May 1959. Portions of the airframe were covered with DX.3, rubber-like anti-radar material, and infra-red suppressors were fitted to the jet pipes.

The Tay-Viscount, VX217, landing at Seighford on 13 January 1958. It had been fitted by Boulton Paul with electric signalling on all three control axes, the first full 'fly-by-wire' aircraft in the world.

The cockpit of the Tay-Viscount. Test pilot Dickie Mancus played a large role in this programme, which was held up for a short while when a robin was found nesting in a wing - and had enjoyed several flights sitting on its eggs!

One of Seighford's hangars crowded with Canberras in January 1960. The one in the foreground is believed to be WV787, fitted with the Buccaneer's 'Blue Parrot' radar. This aircraft is now preserved at the Newark Air Museum.

On the same day this Canberra B.6, WT305, was on the airfield, having had nose modifications for radar trials at CSE Watton.

An aerial view of the hangar complex showing the Seighford to Eccleshall road, which splits the site. The resurfaced runway runs from right to left.

A very poor photograph, but one capturing Canberra PR.7, WH779, making an emergency landing with the wheels half up, on New Year's Day 1959. Boulton Paul Chief Test Pilot, A.E. 'Ben' Gunn had total hydraulic faulire on take-off, and after flying round to burn off fuel, he tried a landing with no flaps, no brakes, no nosewheel and the mainwheels half-up. The landing was reasonable until he hit a huge tree stump in the run-off area. He then retired to the Hollybush in Seighford Village to celebrate the New Year in a more sensible fashion.

Canberra T.4, WH944, in one of the hangars at Seighford in March 1962. Note the side by side ejection seats of the trainer version.

A pre-production Lightning F.1 in a hangar at Seighford as part of the F.3 development programme. Eleven F.1 Lightnings were involved in this programme, each being partially modified.

Another Lightning F.1, XG307, on the airfield. It is fitted with an intake guard. Unlike the Canberras, all the Lightnings were flown by English Electric test pilots.

A very special Lightning at Seighford in June 1960. It is XA847, the P.1B prototype and therefore the very first true Lightning. The special paint scheme on the nose was for the type's naming ceremony. It has been fitted with a mock-up of the F.6 belly-pack, designed and built by Boulton Paul.

Lightning F.3, XG335, fully converted with the larger F.3 fin, at Seighford on 29 March 1962.

The cockpit of the same Lightning, XG335, taken on the same day. It shows what a crowded existence Lightning pilots enjoyed.

One of the last Canberra modifications carried out by Boulton Paul was the Nord AS.30 missile installation. This is Canberra B.15, WH967, with AS.30s on the outboard pylons and rocket pods on the inners.

The last Canberra out of Seighford. Boulton Paul's last aircraft was T.4, WH844, which left the airfield on 23 December 1965. This is T.4, WH840, with a pair of Lightning wings alongside, part of a failed attempt to establish an aircraft museum there. In 1994 it left for the Norfolk and Suffolk Aircraft Museum at Flixton.

Eleven

RAF Tatenhill

Lost down the lanes of eastern Staffordshire, and a long way even from the village which gave it its name, is the surprisingly large airfield which began life as RAF Tatenhill. Built as a satellite to RAF Lichfield, a flight of No.27 OTU Wellingtons arrived in November 1941. In the event it was decided that Tatenhill was unsuitable for a Wellington unit, and it left in October the following year.

Tatenhill then became a satellite for the Oxfords of No.15 (P) AFU for a while, until they too were replaced by the Miles Masters of No.5 (P) AFU from Tern Hill. They stayed until January 1944, when Tatenhill became one of several satellite airfields for No.21 (P) AFU at Wheaton Aston.

Just three miles north of the airfield was the largest underground explosives depot in the country at Fauld. On 22 November 1944 this blew up, in the biggest explosion ever heard in this country. Number 21 MU, the resident unit, or what was left of it, needed a new home and they took over Tatenhill from No.21 (P) AFU, to be joined by the RAF School of Explosives. The airfield was sold after the Second World War and largely returned to agriculture, with the hangars being pulled down.

When Bass's Breweries at nearby Burton needed somewhere to site their corporate aircraft Tatenhill was the obvious place. Half of a standard Bellman hangar was erected and covered with fibre-glass for insulation. A control tower was built into a corner of the hangar and slowly, a thriving general aviation airfield has built up, with a resident aircraft population which includes a civil Jet Provost. Visitors as large as a Norwegian Air Force Hercules have been seen.

A No.27 OTU Wellington B.X, LN710, dropping a bomb probably on a range. Tatenhill was a satellite for No.27 OTU at Lichfield for almost a year, but it is not known from which airfield this aircraft took off.

An aerial view of post-war Tatenhill looking roughly north, showing the main runway with the single hangar and airfield buildings to the right.

The cars rather give away the fact that this is not a wartime scene, but is in fact a civilian-owned de Havilland Queen Bee in front of Tatenhill's post-war tower.

The appropriately registered North American AT-6C Harvard, G-TSIX, taxiing from the runway at Tatenhill where it was rebuilt. It had served with the Portugese and South African Air Forces. The aircraft is now based at Breighton.

The Harvard's stablemate, Mustang XL954, on its annual Christmas trip to Tatenhill to give the owner's father a flight. On the wing is Mike Shelton, Tatenhill Aviation's Chief Engineer.

A privately-owned Percival Pembroke, XL954, about to take-off at Tatenhill, where it was based for a while. It was registered in America as N4234C, but was never exported and now resides with Air Atlantique at Coventry.

The Isaacs Spitfire, G-BBJI, a half-scale homebuilt, at one of the annual PFA Fly-ins at Tatenhill. It is a reminder that a few Spitfires were based at Tatenhill towards the end of the Second World War, providing refresher flying for newly trained pilots who had been sent home because of an aircrew surplus.

The view from the tower at Tatenhill with the engineless Piper Aztec, G-ATFF, in the foreground.

The former crop-sprayer, Piper Pawnee, G-CMGC, which was rebuilt at Tatenhill after a crash, about to make its first post-rebuild flight at the hands of Mike Shelton. It is now used to tow the gliders of the Marchington Gliding Club at Tatenhill.

The McAlpine Aviation-operated Britten-Norman Islander, G-ORED, used by the Red Devils parachute team. It was based at Tatenhill for a while.

The Cessna Citation, G-OMCL, in use by Carlton TV, on a visit to Tatenhilll, which attracts most of the corporate aircraft flights to the northern half of Staffordshire.

Twelve

RAF Wheaton Aston

Built to the north of Wheaton Aston and known locally as Little Onn, after the nearest hamlet, RAF Wheaton Aston was built in 1941 as a substantial satellite of No.30 OTU at Hixon, with three tarmac runways, three large hangars and eight smaller ones. When it opened in December 1941 it was transferred to No.11 Service Flying Training School at Shawbury, where runways were under construction.

The Airspeed Oxfords of No.11 SFTS moved in, and the Oxford was to be the main aircraft associated with Wheaton Aston. The unit was redesignated No.11 (Pilot) Advanced Flying Unit, and was mainly engaged in re-educating Dominion and American-trained pilots into the beauty of flying in British weather conditions.

It became such a large unit that it was split in two and No.21 (P)AFU was created from it, based at Wheaton Aston with Perton as its satellite. An initial complement of seventy-one Oxfords steadily grew until there were nearly 148, and Tatenhill was added as another satellite. The local sport in the potato picking season was trying to hit one of the swarming Oxfords with a well aimed spud !

For a short while it was said that Wheaton Aston was the busiest airfield in the country, but flying began running down as the end of the Second World War approached. The AFU moved out on 1 December 1946, and the station closed on 31 July 1947.

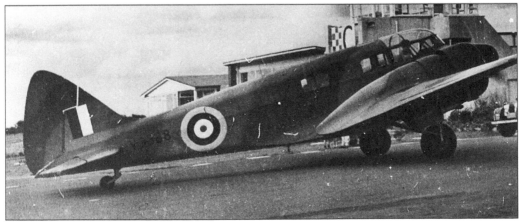

An Airspeed Oxford, and although not one of the many based at Wheaton Aston, it was one seen above the airfield for longer than any other. It is G-AHTW (V3388), Boulton Paul's corporate aircraft which often made the trip from Wolverhampton to their Flight Test Centre at Seighford.

An aerial view of RAF Wheaton Aston around the end of the Second World War. At least forty-six aircraft can be seen parked around the airfield. The main hangar complex is centre right, but there are many aircraft parked round the two blisters north of that and on the other side of the runways.

Pilot Officer Dennis Kyne of the Royal New Zealand Air Force. On 19 October 1943 he took off from Wheaton Aston for his first night cross-country flight to Condover, Shawbury, and back to base. Sometime during the night, in rapidly deteriorating weather, his Oxford, LX518, crashed into Margery Hill in the Peak District. Way off course, his body, lying in the wreckage, was not discovered for four days.

The interior of an Airspeed Oxford. The co-pilots seat has been drawn back, probably to make room for a navigator's chart table.

Also in strength with No.21 (P) AFU were four Avro Ansons including this one, K8723, but this picture was taken at Tern Hill when the aircraft was serving with No.10 FTS. After repair it served four other units before arriving at its last, No.21 (P) AFU at Wheaton Aston. It was finally struck off charge in May 1944.

A famous picture of a Boulton Paul Defiant on flight test from Wolverhampton, over Belvide Reservoir on the A5. It was used on Boulton Paul's Christmas Card in 1940, and was sold as a postcard. The site of RAF Wheaton Aston is the light area just in front of the fin.

Thirteen

Other Airfields

Just what is an airfield? A fully licenced and staffed operation, or just a field where a farmer keeps his aircraft? Each of the first twelve sections of this book features an important airfield in the history of aviation in Staffordshire, but there were others which ought to be mentioned.

There was the strip at Halford's Lane, Smethwick where First World War bombers built by the Birmingham & Midland Carriage Co. first flew. In the Second World War there were relief landing grounds at Penkridge, Abbots Bromley and Battlestead Hill, with busy circuits full of Tiger Moths and Magisters, and there were airstrips cut into the parkland of stately homes at Hoar Cross and Teddesley Park which were used for aircraft storage. Despite much effort and searching, no photographs of these airfields have emerged apart from Penkridge, which the Staffordshire Flying Club has partially opened as a microlight centre, renamed Otherton.

Another storage strip was at Weston Park, but though the stately home is in Staffordshire, the strip was in Shropshire! Nevertheless, recent air displays at Weston Park have been on the Staffordshire side of the estate - but is this really an airfield, even though flying takes place there regularly? Nothing is simple. The most important RAF station in the county is RAF Stafford, but that does not have an airfield! However, RAF Stafford is included in this volume because of its importance and because aircraft have always been present there.

The newest airfield in the county is Roddige, the home of the microlights of Fairflight Aviation, but they still consider themselves to be at Lichfield, whatever the CAA calls it. Amazingly, they are only one field away from a farmer's strip and used his windsock for guidance when their own blew away in a gale!

Handley Page O/400, D5440, with wings folded at the Halford Lane strip, Smethwick, in September 1918. One of 104 O/400s built by the Birmingham & Midland Carriage Co., they were followed by 100 DH.10 Amiens medium bombers.

Amid the wheatfields to the east of Stafford on 16 August 1939, the RAF's huge new storage base begins to take shape.

The foundations of the headquarters building, swarming with bricklayers. RAF Stafford was built by Alfred McAlpine & Son Ltd, following hard on their first large contract to build RAF Cosford just down the road.

Guarding the entrance of RAF Stafford in the 1960s, was this Gloster Javelin FAW.2, XA801 (7739M), which was 'F' of No.46 Squadron. It was sold for scrap in 1993 and replaced by a Harrier.

Outside the headquarters building were a Bristol Bloodhound missile and Javelin FAW.8, XH980, ex-41 Squadron. The Javelin later moved to West Raynham and when that base was closed, it too went for scrap.

An RAF Gazelle, XZ937, at Weston Park Air Day in July 1986. The trouble with helicopters is that everywhere becomes an airfield. There was a relief landing ground at Weston Park during the Second World War where 9 MU at Cosford stored aircraft, but that was on the Shropshire part of the estate; the annual Air Day takes place across the border in Staffordshire.

To illustrate the last point, here a Royal Flight Wessex helicopter turns Molineux football ground, Wolverhampton, into an 'airfield'. As Wolves have always employed the long-ball game, perhaps this is an apt description.

An aerial view of RAF Penkridge taken in 1993, showing the central square of the current microlight field, renamed Otherton. All the airfield buildings, including five small hangars, occupied in the Second World War by No.28 EFTS from Wolverhampton, are to the left of the picture, now part of Pillaton Farm.

A Gemini flex-wing microlight, with the author in the back seat, about to take off from Otherton, the home of the South Staffordshire Aero Club. The airfield buildings are in the background. Substitute a Tiger Moth for the microlight and the scene could be wartime.

A Microflight Spectrum, G-MVXH, at Roddige, just the other side of the A38 from Fradley, and opened by Ultraflight Aviation. Compare this with the first aircraft in the book: similar configuration, similar size, similar row of hangars behind. Only the materials used and the efficiency of the engine seem to have changed in the intervening eighty-seven years.

Acknowledgements

I have to thank a number of people and organisations for lending me photographs for use in this book, but in particular the West Midlands Aviation Archive of the Boulton Paul Association; Maurice Marsh and Dave Welch for numerous pictures of Walsall and Wolverhampton; Derek and Angela Smith of the RAF Lichfield Association; Mike Ruttle for pictures of Stoke and Lichfield; Grahame Beale of Bobbington Estates Ltd; the *Burton Daily Mail* and the Graham Nutt Collection for Bass's Meadow, Burton; Mrs V. Clarkson, C.G. Jones, R.W. Nowell, Eric Ralphs and Keith Rogers all for pictures of Stoke; Grp Capt Edwin Shipley for photos from his excellent book *Walsall Aviation*; the Princess Irene Brigade of the Dutch Army for pictures of RAF Perton, and also Barry Abraham, Jim Boulton, Peter Brew, Jack Chambers, Kevin Dan-Beeching, Mrs Les Freeman, Eric Holden, Jack Holmes, Grahame Hughes, Albert Littleford, Chris Martin, Tom Renshaw, Keith Sedgewick, Andy and Ray Simpson, Andy Thomas and not least, a long-suffering Wendy Matthiason.